I0446893

Table of Contents

To my beloved mom, Marianne,
Your unwavering strength, unconditional love, and quiet resilience have nurtured me into the person I am today. Through both joyous celebrations and turbulent storms, your gentle wisdom has been my guiding light.
You taught me to pursue my passions relentlessly, even when the path was unclear. You showed me how to lift up those around me through selfless compassion. And you ingrained in me the integrity to stay true to my values, no matter the circumstance.
I cannot fully express my eternal gratitude for the person you helped shape me to become. This book, and any good I can put into the world, is thanks to your profound influence.

Note from Author

As world leaders gathered to discuss climate change at COP28, it became clear that impactful action remains elusive. Our leaders are unlikely to implement meaningful policies to mitigate climate change. To spur progress, we must examine why and how our buildings contribute so heavily to emissions.

This book does not claim to solve the climate crisis or provide technical specifications. It simply outlines basics around building decarbonization, making concepts accessible from boiler rooms to boardrooms. Hopefully it inspires you to learn about buildings in your own community and workplace, and empowers you to ask "why is that designed this way?" and "how can it improve?"

Buildings offer immense potential for rapid emissions cuts, producing nearly 40% of global CO_2. Transforming them promises a huge climate impact. However, financial, behavioral and political roadblocks constrain this opportunity, demanding cooperation to navigate them.

This book aims to accelerate building decarbonization by explaining actionable pathways for emissions reductions. It explores solutions around efficiency, electrification, renewables, materials and more. Real-world examples demonstrate outcomes from pioneering low-carbon projects. Critically, it stresses that equitable processes enable just outcomes, analyzing policy and investment through a social justice lens.

Though inertia and upfront costs pose barriers, clean technology trends and ethical investment signal momentum for change. Ambitious codes and standards plans foreground affordability and resilience. Cooperation around open-source roadmaps breeds optimism.

Ultimately, buildings sit at the intersection of climate action and human need, shaping communities and lived experiences. Their integrated transition supports sustainability and wellbeing. By collectively confronting obstacles as stakeholders, our shared future relies on accelerating progress towards buildings that serve both people and the planet.

Greenhouse Gas Emissions from Buildings

Greenhouse gasses (GHGs) are gasses in the Earth's atmosphere that trap heat. The chemical structure of these gasses allow sunlight to enter the atmosphere freely. However, when sunlight is reflected back into space, these gasses trap some of the outgoing energy, retaining heat somewhat like the glass panels of a greenhouse.

The primary GHGs in the Earth's atmosphere are water vapor, carbon dioxide (CO_2), methane (CH_4), nitrous oxide (N_2O), and fluorinated gasses. Buildings play a significant role in global GHG emissions. According to the International Energy Agency's (IEA) 2019 report buildings and their construction together account for 36 percent of global energy use and 39 percent of energy-related carbon dioxide emissions annually.[1] The percentage is due to factors such as the energy consumed in building construction, the use of fossil fuels in buildings for heating, cooling, and lighting, and the energy consumed in the production of building materials. Understanding the lifecycle of GHG emissions in buildings is crucial for developing effective strategies for building decarbonization. This lifecycle can be divided into three main stages:

Construction: This stage includes the extraction and processing of raw materials, manufacturing of building materials, and the construction process itself. Each of these activities involves energy consumption and GHG emissions.

Operation: This is the longest stage in the lifecycle of a building and typically involves the highest GHG emissions. Energy is consumed for heating, cooling, lighting, and use of appliances and equipment. The energy source (e.g., electricity, natural gas, oil) and the efficiency of the building and its systems significantly impact the level of GHG emissions.

Demolition: At the end of a building's life, the demolition process and the disposal or recycling of building materials can also contribute to GHG emissions.

---

[1] IEA (2019), Global Status Report for Buildings and Construction 2019, IEA, Paris.https://www.iea.org/reports/global-status-report-for-buildings-and-construction-2019, License: CC BY 4.0

Conceptualizing Building Decarbonization

Building decarbonization refers to the process of reducing or completely eliminating carbon dioxide emissions caused by energy use in buildings[2]. It involves the implementation of various strategies aimed at minimizing the carbon footprint of buildings, from their design and construction to their operation and maintenance.

The importance of building decarbonization cannot be overstated. Buildings account for nearly 40% of global energy-related carbon dioxide emissions, making them a significant contributor to climate change. By decarbonizing buildings, we can substantially reduce global greenhouse gas emissions and mitigate the impacts of climate change.

Building decarbonization plays a crucial role in combating climate change. As mentioned earlier, buildings are a major source of carbon dioxide emissions. By reducing or eliminating these emissions, we can significantly slow down the rate of global warming. Moreover, building decarbonization can also help us adapt to the effects of climate change. For instance, energy-efficient buildings are generally more comfortable and healthier to live in, thanks to better indoor air quality and temperature regulation. They are also more resilient to extreme weather events, which are expected to become more frequent and severe due to climate change.

Energy efficiency is a key component of building decarbonization. The more energy-efficient a building is, the less energy it needs to operate, and the fewer carbon emissions it produces. Energy efficiency is achievable through a variety of options including: using high-quality insulation, installing energy-efficient appliances and lighting, and designing buildings to take advantage of natural light and ventilation. These options help to lower emissions and save costs. However, energy efficiency alone is not enough to fully decarbonize buildings. We also need to transition to renewable energy sources, such as solar and wind, and implement carbon capture and storage technologies. By combining energy efficiency with these other strategies, we can achieve true building decarbonization and make a significant contribution to the fight against climate change.

---

[2] California Energy Commission (2021), Assembly Bill 3232 and the California Building Decarbonization Assessment, CA, Sacramento https://www.energy.ca.gov/sites/default/files/2021-08/AB3232_Building_Decarbonization_Assessment_Factsheet_ADA.pdf

Building Materials Impact on GHG Emissions

Building materials are a significant contributor to global greenhouse gas (GHG) emissions. From the extraction of raw materials to the manufacturing and transportation processes, each stage of a material's life cycle contributes to its overall carbon footprint. Building materials are a significant contributor to global greenhouse gas (GHG) emissions. From the extraction of raw materials to the manufacturing and transportation processes, each stage of a material's life cycle contributes to its overall carbon footprint.

The expected increase in infrastructure development worldwide could lead building material emissions to double by 2050 if solutions for decarbonization are not implemented on a wide scale across the industry. Common building materials and their emissions include:

Concrete: Concrete is one of the most widely used building materials globally. However, its production process is energy-intensive and releases a significant amount of CO2. Approximately 8% of global CO2 emissions are attributed to concrete production.[3]

Steel: Steel is another commonly used material in construction. Its production involves coal, which is a major source of CO2 emissions and steel accounts for nearly 11% of global CO2 emissions.[4]

Brick: Traditional brick manufacturing involves firing clay at high temperatures, which requires a lot of energy and results in significant CO2 emissions.

Wood: While wood is a renewable resource, its use in construction still contributes to GHG emissions, primarily due to deforestation and the energy used in processing and transportation.

---

[3] Tutton, Mark. "Concrete Is a Huge Source of Carbon Emissions. These Researchers Are Working to Make It Greener." *CNN*, Cable News Network, 23 June 2023, www.cnn.com/2023/06/16/world/concrete-carbon-emissions-researchers-working-to-make-it-greener-climate-scn-spc/index.html.

[4] *Steel climate impact - an international benchmarking of energy and CO2 intensities*. Global Efficiency Intelligence. (2023). https://www.globalefficiencyintel.com/steel-climate-impact-international-benchmarking-energy-co2-intensities#:~:text=The%20iron%20and%20steel%20industry,carbon%20dioxide%20(CO2)%20emissions.

This is typically known as, embodied carbon, which refers to the amount of carbon dioxide (CO_2) emitted during the production, transportation, and installation of a building material. It's a crucial aspect of a material's overall carbon footprint. For instance, while a wood-framed building might have lower operational emissions than a concrete one, the embodied carbon from deforestation could make the wood building's total carbon footprint higher. Therefore, it's important to consider both embodied and operational carbon when assessing a building's environmental impact. There are several strategies to reduce the carbon footprint of building materials: including

Material Efficiency: This involves using less material by optimizing design and construction techniques. For instance, using advanced framing techniques can reduce the amount of wood used in a building.

Material Substitution: This involves replacing high-carbon materials with low-carbon alternatives. For example, using engineered wood products instead of concrete or steel can significantly reduce a building's embodied carbon.

Recycled Materials: Using recycled or reclaimed materials can also reduce a building's carbon footprint. This not only reduces the need for new material production but also diverts waste from landfills.

Local Sourcing: Sourcing materials locally can reduce emissions from transportation. It also supports local economies and can often result in a building that is more in tune with its local environment.

The Impact of Building Methods and Systems on GHG Emissions

Building methods and systems refer to the various techniques and processes used in the construction and operation of buildings. These include the choice of building materials, construction techniques, heating, ventilation, and air conditioning (HVAC) systems, lighting systems, water systems, and waste management systems. Traditional building methods often involve the use of energy-intensive materials and processes and are major sources of greenhouse gas emissions.

The carbon footprint of a building method or system is the total amount of greenhouse gas emissions produced over its lifecycle, from the extraction and processing of raw materials to construction, operation, and eventual demolition or deconstruction. Concrete and steel, as mentioned earlier, have a high carbon footprint due to the energy intensive processes involved in their production. The use of these materials in construction, therefore, contributes significantly to the carbon footprint of buildings.

HVAC systems also have a substantial carbon footprint. The burning of fossil fuels for heating releases CO_2, while the electricity used for cooling often comes from power plants that burn fossil fuels. Moreover, many HVAC systems use refrigerants that are potent greenhouse gasses.

Lighting systems, particularly those that use incandescent or halogen bulbs, can also contribute to a building's carbon footprint. These types of bulbs are energy-inefficient, meaning that a significant portion of the electricity they consume is wasted as heat.

There are several strategies that can be employed to reduce the carbon footprint of building methods and systems including:

Using low-carbon building materials: This could involve using recycled or reclaimed materials, as well as materials that require less energy to produce. For instance, wood is a renewable resource that absorbs CO_2 as it grows, making it a lower carbon alternative to concrete and steel.

Improving energy efficiency: This can be achieved by using high-efficiency HVAC systems, LED lighting, and energy-efficient appliances. Insulation can also play a crucial role in reducing energy consumption for heating and cooling.

Implementing renewable energy systems: Solar panels, wind turbines, and geothermal systems can provide clean, renewable energy for buildings, reducing the need for fossil fuels.

Designing for longevity and adaptability: Buildings that are designed to last and to be easily adapted for different uses can reduce the need for new construction and the associated emissions.

Decarbonization Strategies for Buildings

Building decarbonization is a multi-faceted approach that involves several strategies. These include the use of renewable energy sources, improving energy efficiency, using low-carbon building materials, and implementing sustainable building design principles.

Renewable Energy Sources: These are energy sources that are naturally replenishing such as solar, wind, and geothermal energy. By replacing fossil fuels with renewable energy sources, we can significantly reduce the carbon emissions associated with building operations. Solar panels, wind turbines, and geothermal systems are some of the renewable technologies that can be integrated into buildings. These technologies not only reduce carbon emissions but also provide a host of other benefits such as reduced energy costs and improved energy security.

Energy Efficiency: This involves reducing the amount of energy required to provide the same level of service. It can be achieved through various means such as using energy-efficient equipment, improving insulation, and optimizing building design. By reducing the amount of energy a building needs, we can reduce the amount of carbon emissions associated with its operation. Energy efficiency can be improved in a number of ways. This includes using energy efficient appliances, improving insulation, and using energy management systems. Energy-efficient buildings are not only good for the environment but also for the occupants. They provide a more comfortable living environment and can significantly reduce energy bills.

Low-Carbon Building Materials: The use of low-carbon building materials can significantly reduce the carbon footprint of a building. These materials include recycled or sustainably sourced timber, low-carbon concrete, and other innovative materials.

Sustainable Building Design: This involves designing buildings in a way that minimizes their environmental impact. This can be achieved by optimizing the building orientation, using passive design principles, and integrating green spaces.
Policies and Regulations Related to Building Decarbonization
Building decarbonization is not just a technical challenge but also a policy issue. Policies and regulations play a crucial role in driving the decarbonization of buildings. We will cover more of this in the later chapter entitled 'The Role of Policy and Regulation.'

The Future of Building Decarbonization

Building decarbonization is not just a current trend, but it is the future of the construction and real estate industry[5]. There are some promising trends that seem to be directing innovation in this space. For example, construction companies nowadays are finding clever ways of using recycled and eco-conscious materials like repurposed steel, bio-materials and cleaner concrete mixes. Integrating these recycled goods significantly lightens the carbon footprint left behind from the building process. Additionally, renewable power sources like solar and wind are being seamlessly incorporated into structures to eliminate emissions and high energy bills altogether.

Pioneering new concepts and strategies is integral for pushing building decarbonization forward. Novel technologies that bolster efficiency, coupled with breakthroughs in renewables that facilitate energy capture, are shrinking the carbon profile of modern buildings. Insulation techniques are becoming incredibly effective at sealing in heat, while innovations in solar panels and wind turbines are making self-sustainability more turnkey.

Experimentation with construction approaches themselves are also gaining traction - passive home architecture that strategically harnesses sunlight, insulation and airflow is an example of a model optimized for energy conservation. As priorities expand to consider environmental impact, the techniques used to erect cleaner structures will likely continue to be inventive and multi-faceted. There seem to be some encouraging indicators that carbon-conscious trends are beginning to transform the future of how buildings are conceived and fabricated.

Innovations in sustainable technology are prompting policy makers to consider new regulations that encourage cleaner construction standards. Implementing decarbonization frameworks in the building sector could be integral for curbing emissions and promoting better environmental stewardship. Beyond benefiting the planet, transforming modern infrastructure to be more carbon-neutral also carries tangible economic and social perks. Energy efficient buildings can drastically cut costs for owners and occupants while creating jobs for those specializing in renewable tech and eco-conscious design. Stimulating the growth of these emerging green industries could ripple through the economy. On an individual

---

[5] "Call for Action: Seizing the Decarbonization Opportunity in Construction." *McKinsey & Company*, McKinsey & Company, 14 July 2021, www.mckinsey.com/industries/engineering-construction-and-building-materials/our-insights/call-for-action-seizing-the-decarbonization-opportunity-in-construction.

level, improving indoor air quality and energy efficiency can pay dividends as well by creating healthier, more comfortable spaces to inhabit. As climate consciousness becomes more embedded in contemporary values, forward-thinking policies that champion decarbonized buildings have the potential to fundamentally transform what the infrastructure of the future looks like. The 21st century could witness sweeping changes in best practices for sustainable design and clean energy integration.

Energy Efficiency

Energy efficiency, in the context of buildings, refers to the use of technology and practices to reduce the amount of energy required to provide products and services within the building. This could involve anything from using energy-efficient appliances and lighting to implementing advanced building designs and materials that minimize energy loss.

Energy efficiency in buildings is paramount as buildings account for nearly 40% of global energy-related CO_2 emissions, making them a significant contributor to climate change.[6] By improving energy efficiency, we can reduce this impact, saving energy and reducing greenhouse gas emissions. Moreover, energy efficiency can also lead to significant cost savings over time, making it not only an environmentally responsible choice but also a financially sound one.

Insulation

Insulation plays a crucial role in the energy efficiency of buildings. It acts as a barrier to heat flow, reducing the amount of heat that escapes from a building in the winter and preventing heat from entering the building in the summer. This reduces the need for heating and cooling systems, saving energy and reducing greenhouse gas emissions.

Insulation can be installed in various parts of a building, including the roof, walls, and floors. The effectiveness of insulation is measured by its R-value, which indicates its resistance to heat flow. The higher the R-value, the more effective the insulation. There are several types of standard insulation materials available, each with its own advantages and disadvantages:

Fiberglass: This is the most common type of insulation. It is made from fine glass fibers and is often used in batts and rolls. It has a moderate R-value and is relatively inexpensive. However, the small glass fibers can irritate the skin, eyes, and lungs.

Mineral Wool: This includes rock wool and slag wool. It is made from molten glass, stone, or industrial waste that is spun into fibers. It has a higher R-value than

---

[6] *How Green Buildings Can Help Fight Climate Change | U.S. Green Building Council.* 1 Mar. 2021, www.usgbc.org/articles/how-green-buildings-can-help-fight-climate-change.

fiberglass and is fire resistant. The fibers can cause skin, eye, and respiratory irritation.

Cellulose: This is made from recycled paper that is treated with fire retardants. It has a higher R-value than fiberglass. The dust created during installation can cause respiratory issues if inhaled.

Polyurethane Foam: This is a type of spray foam insulation that expands when applied, filling gaps and cracks. It has a high R-value and provides excellent air sealing. The chemicals released during installation can cause breathing and skin irritation.

Polystyrene: This includes expanded polystyrene (EPS) and extruded polystyrene (XPS). These are rigid foam boards that have a high R-value and are often used for insulating foundations and roofs. The manufacturing process uses a blowing agent that can contribute to smog and climate change.

The choice of insulation material will depend on a variety of factors, including the climate, the type of building, and the part of the building where the insulation will be installed. It is important to choose a material with an appropriate R-value for the specific application. Installation is also crucial to the effectiveness of insulation. Even the best insulation material will not perform effectively if it is not installed properly. It is important to ensure that the insulation is evenly distributed, with no gaps or compression, as this can significantly reduce its effectiveness.

Professional installation is recommended for most types of insulation, especially spray foam and rigid foam board. However, some types of insulation, such as fiberglass batts, can be installed by homeowners with some DIY experience. By understanding the different types of insulation materials and their effectiveness, and by ensuring proper installation, building professionals can significantly reduce the energy consumption and carbon footprint of buildings. Insulation products should be thoroughly researched to understand the potential health risks related to the installation and lifecycle of the product.

Appliances

In any building, be it residential or commercial, appliances account for a significant portion of energy consumption. This includes everything from refrigerators and air conditioners to computers and lighting fixtures. Each of these appliances consumes energy to function, and depending on their efficiency, this consumption can vary significantly. Therefore, it's essential to understand the role of appliances in a building's overall energy consumption to effectively reduce its carbon footprint. Energy-efficient appliances are designed to use less energy for the same or improved level of service compared to traditional models. These appliances come in various types, including energy-efficient refrigerators, air conditioners, heaters, washing machines, dishwashers, and light bulbs, among others.

The benefits of energy-efficient appliances are manifold. They not only help reduce a building's energy consumption and thus its carbon footprint, but they can also result in significant cost savings over time due to reduced energy bills.

When choosing energy-efficient appliances, consider the following factors:

Energy Efficiency Rating: Look for appliances with a high energy efficiency rating. The higher the rating, the more efficient the appliance is likely to be.

Size: Choose the size of the appliance based on your needs. Larger appliances tend to consume more energy, so avoid buying an appliance that is larger than what you need.

Features: Some features can increase an appliance's energy consumption. For example, a refrigerator with a through-the-door ice dispenser typically uses more energy than a model without this feature.

Cost: While energy-efficient appliances may be more expensive upfront, they can save money in the long run through reduced energy bills.

The Energy STAR® Rating System

The Energy STAR® rating system is a widely recognized symbol for energy efficiency. It was established by the U.S. Environmental Protection Agency to help consumers identify and purchase energy-efficient products. Appliances that carry the Energy STAR® label have been independently certified to save energy without sacrificing features or functionality. They use 10-50% less energy than standard appliances and thus can help reduce a building's energy consumption and carbon footprint.

Lighting

Lighting is a significant contributor to a building's energy consumption. It accounts for approximately 20% of the total electricity used in commercial buildings and 10% in residential buildings. Therefore, improving the efficiency of lighting systems can play a crucial role in reducing a building's energy consumption and, consequently, its carbon footprint. Today, the best choice for lighting is installing LEDs.

Light Emitting Diodes (LEDs): LEDs are even more energy-efficient than CFLs and have a longer lifespan. They use about 75% less energy than incandescent bulbs and can last up to 25 times longer.

When it comes to installation, energy-efficient bulbs can often be used in existing light fixtures. However, some fixtures may need to be replaced to accommodate certain types of energy efficient bulbs. It's always best to consult with a lighting professional or the bulb manufacturer for specific installation instructions.

Transitioning to energy-efficient lighting is a simple yet effective strategy for reducing a building's energy consumption and carbon footprint. By understanding the different types of energy-efficient lighting technologies and how to choose and install them, building professionals can make informed decisions that contribute to building decarbonization.

Power Over Ethernet

Power distribution is another major energy user in buildings. In commercial buildings, office equipment and devices can account for up to 20% of electricity consumption. As such, employing efficient methods of power delivery could notably cut electricity usage and carbon emissions from buildings. One promising technology is power over ethernet (PoE).

Power Over Ethernet: PoE allows electric power to be transmitted over existing ethernet cable infrastructure. This removes the need for separate power cords and outlets for each device.
PoE-enabled switches and controllers can deliver power through the same CAT 5 or CAT 6 cables used for data transmission.

PoE technology can utilize existing ethernet infrastructure already in place in many buildings. However, upgraded PoE switches and injectors may be required to handle the higher power capacity. Proper installation by qualified professionals is important for both safety and reliability. Still, switching to PoE for power delivery can be less invasive and costly than running entirely new electrical lines in a building. By cutting extraneous cabling and enabling more precise power control, PoE both improves efficiency and energy monitoring. The technology is versatile enough for applications from VoIP phones to security cameras to lighting. As energy costs rise, PoE's ability to optimize power usage makes it an increasingly attractive avenue for supporting building decarbonization through reduced consumption.

Additionally, PoE technology offers reliability and resilience benefits for critical systems. By transmitting both power and data over a single ethernet cable infrastructure, there are fewer potential failure points compared to separate power and networking lines. The centralized power sourcing also allows easy backup power integration through uninterruptible PoE switches connected to generators or batteries. Being able to maintain continuous uptime for key systems like access control, fire alarms, and emergency lighting improves building resilience against power disturbances like outages or voltage spikes. Especially for vital systems, having power delivery handled through the resilient PoE framework versus conventional AC outlets enhances overall building reliability against disruptive events. As climate change drives more extreme weather causing infrastructure stresses, PoE capabilities for supporting continuity of important functions grows in importance for building-level resilience.

HVAC Systems

HVAC (Heating, Ventilation, and Air Conditioning) systems play a significant role in the energy consumption of buildings. According to the U.S. Energy Information Administration, HVAC systems account for more than 40% of the energy used in commercial buildings in the United States.[7] This is due to the fact that these systems are responsible for maintaining comfortable temperatures and good air quality in buildings. However, traditional HVAC systems can be quite energy-intensive, contributing significantly to a building's carbon footprint. Fortunately, there are several energy-efficient HVAC technologies available today that can significantly reduce a building's energy consumption including:

---

[7] *Use of Energy in Commercial Buildings - U.S. Energy Information Administration (EIA).* www.eia.gov/energyexplained/use-of-energy/commercial-buildings.php.

High-Efficiency Heat Pumps: These devices provide heating and cooling by moving heat from one place to another. They can be two to three times more efficient than traditional electric heating and cooling systems.

Energy Recovery Ventilators (ERVs): ERVs recover energy from exhaust air and use it to precondition incoming fresh air, reducing the energy required for heating or cooling.

Variable Refrigerant Flow (VRF) Systems: VRF systems adjust the flow of refrigerant to different parts of the building based on demand, resulting in more efficient use of energy.

Programmable Thermostats: These devices allow for the setting of temperature schedules, ensuring that the HVAC system is only working when it's needed.

The benefits of these energy-efficient HVAC technologies go beyond just energy savings. They can also improve indoor air quality, enhance occupant comfort, and reduce maintenance costs. Furthermore, they can contribute to building decarbonization by reducing the amount of greenhouse gas emissions associated with building operation. Choosing an energy-efficient HVAC system involves considering factors such as the size and layout of the building, the local climate, and the specific heating and cooling needs of the occupants. It's important to work with a knowledgeable HVAC professional who can provide guidance on the best system for your needs.

Once an energy-efficient HVAC system is installed, regular maintenance is key to ensuring its continued efficiency. This includes tasks such as changing air filters, cleaning coils, and checking system controls. Regular maintenance can also help to extend the lifespan of the system, providing further savings in the long run.

Water Heating

Water heating is a significant energy expense in most buildings, accounting for around 12 percent of total energy consumption. This energy is used for various purposes, such as bathing, cooking, cleaning, and space heating. Traditional water heaters, especially those that use fossil fuels, can contribute significantly to a building's carbon footprint due to their high energy consumption and greenhouse gas emissions. Therefore, improving the energy efficiency of water heating systems is a crucial aspect of building decarbonization. There are several energy-efficient water heating technologies available today, each with its own benefits and considerations. Some of these include:

Heat Pump Water Heaters (HPWHs): These devices use electricity to move heat from the air or ground to heat water, which is much more energy-efficient than traditional electric water heaters. They can reduce water heating energy use by up to 60 percent.

Tankless or Demand-Type Water Heaters: These systems heat water directly without the use of a storage tank, eliminating the standby energy losses associated with storage water heaters. They can be 24-34 percent more energy-efficient than traditional tank water heaters for homes that use a lot of hot water.

Solar Water Heaters: These systems use the sun's energy to heat water. While the upfront costs can be high, they can provide significant energy savings over time, especially in sunny climates.

Condensing Water Heaters: These systems are an option if you heat with gas. They capture extra heat that would otherwise escape from the flue, making them very efficient.

When choosing an energy-efficient water heater, consider the following factors:

Fuel Type and Availability: The fuel type can affect the annual operation costs and the water heater's size and energy efficiency.

Size: To provide your household with enough hot water and to maximize efficiency, you need a properly sized water heater.

Energy Efficiency: While energy-efficient models may seem to cost more initially they often result in significant savings over their lifetime.

Costs: Consider both the purchase cost and the operating cost of the water heater.

Energy Efficiency and Building Design

The influence of building design on energy efficiency is paramount. It's a critical element that shapes a building's energy consumption throughout its lifespan. This process involves careful choices regarding the building's orientation, layout, materials, insulation, and systems, each of which plays a significant role in its overall energy usage.

Building Orientation and Layout Orientation is vital in managing the building's exposure to sunlight, affecting its heating and cooling demands. Layout influences the utilization of natural light, potentially reducing reliance on artificial lighting.

Material and Insulation Choices: Selecting appropriate materials and insulation determines the building's ability to retain heat, impacting the need for additional heating or cooling.

Energy-Efficient Systems: The selection of systems, including HVAC and lighting, is crucial. Incorporating energy-efficient options, like LED lighting and high-efficiency HVAC units, can considerably lower energy consumption.

Renewable Energy Integration: Incorporating renewable energy sources, such as solar panels or geothermal systems, into the design can further enhance a building's energy efficiency.

Sustainable Building Materials:Using energy-efficient materials, like those with high thermal mass, can help regulate indoor temperatures and further reduce energy needs.

Evaluating the Impact of Energy Efficiency Technologies

The effectiveness of energy efficiency in buildings is measurable through various metrics, such as Energy Use Intensity (EUI) and Building Energy Rating (BER). EUI is calculated by dividing the building's annual energy consumption (in kBtu) by its total gross floor area. A lower EUI signifies higher energy efficiency. BER rates a building's energy efficiency from A (most efficient) to G (least efficient), based on expected energy consumption under standard operating conditions.

Energy Modeling Software: This software simulates a building's energy consumption under various scenarios, enabling comparisons of different technologies and design strategies.

Life Cycle Assessment (LCA):LCA evaluates the environmental impact of a material or technology across its lifespan, helping to understand the long-term implications of energy efficiency technologies.

Benchmarking: Comparing your building's energy performance with similar structures provides insights into the effectiveness of energy efficiency technologies.

Energy Audits: Regular energy audits are essential for identifying energy-saving opportunities and assessing the effectiveness of existing measures. These audits involve inspecting the building, analyzing energy bills, identifying opportunities for upgrades, and providing recommendations.

The Future of Energy Efficiency Technologies

Emerging technologies are set to revolutionize building design and operation. These include smart grids, energy storage solutions, smart appliances, and Building Energy Management Systems (BEMS), each contributing uniquely to energy efficiency.

The advancement of these technologies, however, requires a synergy of technological innovation and supportive policy frameworks. Regulations and incentives are crucial for translating these advancements into widespread practice. The combined approach of technology development and policy support is essential for significant strides in reducing buildings' carbon impacts.

AI's Transformative Impact on Building Decarbonization

Artificial Intelligence (AI) stands out as a key technology in enhancing building decarbonization efforts. The emerging technology plays a crucial role in tracking emissions and simulating the impact of equipment upgrades and policy changes, making it an integral part of any comprehensive decarbonization strategy. By utilizing machine learning, big data, and advanced simulations, AI optimizes sustainable building operations in several ways:

Energy Management: AI systems can optimize energy usage for heating, cooling, lighting, etc., based on historical data and real-time external factors, achieving estimated efficiency gains of 10-25%.

Predictive Maintenance: Using machine learning, AI can foresee equipment malfunctions, allowing for energy-efficient maintenance and operations. Renewables Integration: AI algorithms can effectively integrate various renewable energy sources into building operations, supporting the shift away from fossil fuels. **Dynamic Operations:** AI facilitates the adjustment of electricity uses like lighting and elevator operations in real-time, balancing demand with intermittent renewable sources.

Renewable Energy

Renewable energy, often referred to as clean energy, comes from natural sources or processes that are constantly replenished. Unlike fossil fuels, which are inherently dirty and emit harmful greenhouse gasses when burned, renewable energy sources are virtually inexhaustible and environmentally friendly. They include solar, wind, hydro, geothermal, and biomass energy, among others.

Buildings devour over 35% of the planet's energy and spew out greenhouse gasses. Shifting buildings to utilize renewable technologies for power, heating and cooling can significantly shrink energy footprints. Systems that harness solar, geothermal, biofuel and waste heat sources offer sustainable alternatives to fossil fuels reliance. Whether converting sunlight into electricity via panels, channeling subsurface heat with geothermal pumps, or capturing combustible emissions for reuse, renewables tap into clean, freely available energy streams. Adopting these technologies diverts demand away from coal, natural gas and other carbon-intensive resources. The emissions eliminated through renewables integration really starts to add up system-wide.

Beyond environmental perks, integrating renewables often stabilizes energy costs long-term while creating local green tech jobs. The nature of renewables also allows unique architectural integration compared to industrial equipment aesthetics. For example, solar films and panels can overlay curved facades as a second skin. Biosolar roofs sustain plant life while insulating the structure underneath.

Function need not be sacrificed for form either. Renewables can generate electricity, ambient heating/cooling and hot water using free and plentiful resources. The technologies keep advancing in efficiency and application as well. Given their environmental, economic and design benefits, renewables have emerged as a pivotal tool for decarbonizing buildings. And the need to slash emissions across the built world's massive energy footprint grows more urgent by the day. Deploying alternative energy systems provides a path towards an efficiently-powered and lower footprint future.

Solar Energy

Solar Photovoltaic (PV) Systems are a type of renewable energy technology that converts sunlight into electricity. They are composed of several solar cells made of semiconducting materials like silicon, which absorb sunlight and generate an electric current through the photovoltaic effect. Solar PV systems can be installed on rooftops, building facades, or the ground, and can be used in both residential and commercial buildings. They are a popular choice for building decarbonization due to their ability to generate clean, renewable electricity, reduce reliance on fossil fuels, and lower greenhouse gas emissions.

Solar thermal systems, on the other hand, use sunlight to generate heat. They consist of solar collectors (usually flat panels or evacuated tubes) that absorb sunlight and transfer it as heat to a fluid. This heated fluid can then be used to provide hot water or space heating in a building, or it can be used to generate electricity through a heat engine. Solar thermal systems can be an efficient way to reduce the energy consumption of a building, particularly in climates with a high demand for space heating or hot water. They can also contribute to building decarbonization by reducing the use of fossil fuels for heating.

When considering the use of solar energy technologies in a building, it's important to evaluate their feasibility in the specific context of the building. Factors to consider include the building's location, orientation, roof space, roof load capacity, and energy demand, as well as local climate conditions and solar irradiance levels. For example, solar PV systems are most effective in locations with high levels of sunlight and clear skies. They also require a suitable surface for installation, such as a south-facing roof with little or no shading. Solar thermal systems, meanwhile, can be a good option in buildings with a high demand for hot water or space heating, particularly if these demands coincide with periods of high solar irradiance.

Geothermal Energy

Geothermal power comes straight from the Earth's heat, providing a constant supply of clean energy no matter the weather or season. This makes it uniquely reliable compared to other renewables like solar or wind. Specifically, geothermal energy taps into underground reservoirs of hot water and steam, which are heated by the Earth's core. This heat can then be harnessed to reduce the carbon footprint of buildings.

Geothermal heat pumps exemplify this application. They work by pumping fluid through pipes buried underneath buildings. The surrounding soil keeps this ground loop temperate - cooler than the building in summer, warmer in winter. The fluid circulating through it then carries this ambient subsurface heat inside. In essence, the Earth itself serves as a giant heat sink and source.

Crucially, geothermal systems use much less electricity than conventional heating and cooling units. Avoiding fossil fuel combustion also slashes carbon emissions. And unlike solar or wind, the steady subsurface heat offers supply stability, regardless of intermittent clouds or calm days. This makes geothermal energy especially valuable for building decarbonization.

The upfront drilling and installation costs can be high. However, with supportive incentives and programs, geothermal systems can make sense financially through long-term energy savings. And their environmental benefits only compound over years of operation. In short, for both newly built and existing structures, this advanced heating and cooling approach taps into an abundant carbon-free resource right below our feet.

Geothermal energy can be harnessed in various ways. Ground source heat pumps, as mentioned, exchange heat with the shallow subsurface. Deep geothermal plants, on the other hand, tap into deeper high-temperature reservoirs for steam to drive turbines and generate power. Co- and tri-generation facilities combine this heat with biofuels to produce not just electricity but also heating/cooling and hot water.

For new buildings, geo-exchange systems can be designed from the start, with vertical boreholes drilled at the site. Some innovative approaches even integrate geothermal loops into foundation piles needed for construction anyway. For existing buildings, especially in dense urban settings, there are emerging innovations like using ambient heat in sewer and subway lines. This turns infrastructure already below cities into ready geothermal sources.

Crucially, transitioning buildings to geothermal energy aligns with global decarbonization goals. Geothermal technology offers a proven way to sharply reduce heating/cooling emissions from both new and retrofitted buildings. Especially as grids add more renewables, geothermal systems amplify these upstream power emissions reductions through highly efficient operation.

Accelerating their adoption can therefore provide substantial value for climate change mitigation.

Energy Storage

As the world transitions to renewable energy to mitigate climate change, the intermittent nature of sources like solar and wind presents reliability challenges. Energy storage provides a critical solution by capturing temporary surpluses and dispatching this reserve when generation falls short of demand. This tempering capacity delivers consistency without overtaxing carbon-emitting conventional generators. Storage also reduces grid dependence, lowers costs and strengthens resilience. With prudent implementation, it can profoundly complement building-level and grid-wide decarbonization.

Several storage technologies are gaining favor through unique advantages: Battery systems directly convert electricity to chemical energy for later reuse while providing efficiency, flexibility and responsiveness. Lithium ion batteries currently lead the market by combining ample storage density with extended lifespans. Ongoing innovations around battery chemistry aim to balance performance with ethical sourcing and recycling.

Thermal solutions indirectly warehouse energy as heating or cooling capacity for subsequent space conditioning and electricity production. Concrete examples include insulated hot water tanks, chilled water reservoirs and molten salt systems. These assets align electrical and thermal load balancing while easing strain on grids.

Mechanical storage exploits physical forces like gravity and motion to later reconstitute potential or kinetic energy as electricity. Compressed air, pumped hydro and flywheel technologies exemplify this category's versatility and value for large, long-duration applications.

Hydrogen methods electrolyze water to hydrogen gas as an energy carrier when supply exceeds demand. This hydrogen then fuels flexible electricity generation in fuel cells or turbines when required. As production pathways progress, hydrogen promises seasonal storage capacity topping other options.

Tailoring storage to sustainable structures requires weighing numerous factors:

Technical compatibility: considers the local climate, existing infrastructure constraints, site conditions and other context-specific circumstances that determine feasibility.

Cost-effectiveness: balances upfront investments and ongoing operational expenses against potential savings and revenue from utility arbitrage, expanded renewables self-consumption and enhanced resilience. Storage must pencil out financially while meeting sustainability goals.

Evolving regulatory environments: significantly impact motivation and implementation. These include carbon policies, efficiency incentives, renewable and storage procurement mandates, permitting protocols and more.

Within these boundary conditions, appropriately incorporating storage can profoundly expedite building and grid decarbonization through greater energy flexibility, security and optionality. The optimal solutions integrate technological capabilities with economic realities and policy landscapes.

As the building sector pursues rapid decarbonization, energy storage emerges as an indispensable asset. Its capacity to balance intermittent renewables' peaks and troughs while cutting carbon dependence makes it pivotal infrastructure for sustainable energy systems.

Though all storage options have tradeoffs, continuing advances on multiple technology fronts offer expanding possibilities. With costs falling and supportive policies maturing, the path ahead points unambiguously to wide scale storage deployment. Buildings integrating smart storage solutions will lead the next generation of resilient, efficient and ethical structures while forging grid synergies essential to climate progress. The vital layer that storage provides in aligning supply and demand can prove profoundly transformative if applied thoughtfully across linked machines and spaces. Our shared zero-carbon future relies on harvesting storage's versatility through innovative building design and purposeful policy.

Wind Energy

Wind energy harnesses the kinetic power of moving air to generate emissions-free renewable electricity. Wind turbines, positioned to capture robust wind currents, transform this mechanical power into usable energy for direct consumption, storage, or grid transmission. With ample potential supply and rapidly scaling technology, wind promises a pivotal role in building decarbonization and climate progress.

As a distributed or grid-tied resource, wind energy displaces fossil fuels to cut carbon impacts from the building sector. Structures leveraging wind decrease their reliance on greenhouse gas-intensive grid power for heating, cooling and operations. Well-sited turbines can also reduce energy costs and provide resilience co-benefits.

Furthermore, the scalability of wind energy complements growing decarbonization ambitions for zero-energy and energy-positive buildings.
However, feasibility constraints and economic barriers temper universal adoption. Intermittent generation that misaligns with demand profiles requires mitigating measures through storage, load shifting or grid balancing. Turbines must integrate site-specific factors including local wind regimes, height for air access, zoning setbacks and institutional restrictions. Upfront turbine costs can deter investment despite long-term savings and sustainability benefits. Ongoing maintenance and component replacements may also burden owners over system lifetimes spanning decades.

Technology improvements on numerous fronts are expanding possibilities for buildings to harness wind energy. Innovations in distributed wind turbines allow integration with existing structures and infrastructure. Falling component costs are increasing value propositions through better return on investment timelines. Supportive decarbonization policies and incentives are accelerating adoption in regions prioritizing ambitious climate action across the building sector.

The path towards widespread building electrification and decarbonization relies on bold, multiplicative efforts across all clean energy technologies. Wind power represents an increasingly viable and scalable solution within this portfolio, with the potential to make substantial contributions given prudent implementation.

Buildings leveraging appropriately sited wind energy systems can target deeper emissions cuts through direct renewable generation while stimulating broader infrastructure transitions through synergistic grid linkages.

Biomass Energy

Biomass energy harnesses the chemical potential locked within organic materials to generate renewable biofuels, heat and power. Waste streams, dedicated crops, and forestry byproducts containing complex carbon compounds fuel this process through combustion or thermochemical and biochemical conversions. The resulting electricity, steam and liquid fuels displace fossil energy for buildings and infrastructure. With prudent implementation, biomass offers significant decarbonization potential.

As a versatile renewable resource, biomass fuels various heating applications from individual buildings to district systems. Standalone stoves and boilers integrate into existing distribution infrastructure with marginal retrofitting. By substituting wood pellets, chips or agricultural residues for natural gas, these technologies directly reduce carbon emissions from space and water heating. The renewable nature of biomass, along with the potential to offset lifecycle impacts through crop selections and land use choices, enhances the climate value proposition.However, several key factors determine the feasibility and adoption of biogenic heating solutions:

Feedstock availability and supply chain viability affect reliable access within reasonable proximity. Transportation distances and logistics may constrain cost-competitiveness. Storage space for bulk volumes of fuel must not overly burden site capacity.

Upfront retrofit complexity and ongoing operability challenges may hinder uptake. Lack of institutional experience in handling and burning biomass introduces uncertainty. Maintenance requirements for fuel preparation, loading and ash removal deter some adopters.
Full lifecycle emissions depend heavily on source material and cultivation methods. Intensive use of fertilizers, conversion inefficiencies and soil disturbance could diminish net gains. Explicit tracking and disclosure of impacts aids transparency.

Overall, appropriately incorporating biogenic heating and cogeneration to supplement decarbonized grids promises major emissions reductions from the built environment. Realizing ambitious climate goals requires scaling renewables across all viable technology pathways. With conscientious implementation minimizing tradeoffs, biomass energy constitutes a flexible option within this broader portfolio.

Hydro Energy

Hydro energy, also known as hydropower, is a form of renewable energy that harnesses the power of Hydropower converts the innate potential and kinetic energy within flowing water into zero-emission renewable electricity. Typically harnessed through dams and reservoir systems at the grid scale, small run-of-river setups also provide distributed generation options. As water travels from higher to lower elevations, gravity-induced motion spins hydroturbines to activate generators. This simple principle aligns elegantly with building decarbonization by substituting clean electricity to displace carbon-intensive power usage.

Properly sited hydro energy excels at offering consistent and controllable generation capacity with minimal lifecycle impacts. River-adjacent buildings benefit greatly by integrating run-of-river systems that divert a portion of flow without dams or reservoirs. By directly utilizing such small-scale hydropower, structures decrease their reliance upon emission-heavy grid energy. Hydroelectricity can fully electrify or supplement building needs to significantly lower operational carbon footprints. Excess generation may also be sold or stored locally to further improve project economics.

However, hydro energy integration faces barriers around site-specific resources and system requirements. Not all locations boast the vertical drops or minimum flows essential for hydropower viability. Even suitable sites may suffer seasonal variability that necessitates supplemental sources. Turbine configuration for optimal output relies on matching equipment capabilities with available flow energy. Ongoing debris management and component maintenance may increase costs.

Still, the potential hydroelectricity benefits are too substantial to ignore in the building decarbonization toolkit. Hydro stands distinct from solar and wind alternatives with flexible dispatchability immune from intermittency challenges.

This makes it a firming asset able to balance variable generation, especially if reservoir storage is available. Unlocking hydropower co-benefits requires selecting appropriate sites through detailed resource analysis. But once implemented thoughtfully, it constitutes reliable and renewable energy with lifespans typically exceeding half a century. With apt small-scale adoption, hydro energy remains poised to deliver material progress towards deep building decarbonization. The feasibility of using hydro energy in a building context depends on several factors:

Location: The building must be located near a water source with a sufficient flow rate and drop in elevation. This could be a river, stream, or even a man-made waterway.

Regulations: Some regions have strict regulations regarding the use of waterways for power generation. It is important to check local regulations and obtain the necessary permits before installing a hydro system.

Cost: While the operational costs of a hydro system are relatively low, the upfront costs can be significant. However, these can often be offset by the long-term savings on energy bills, as well as potential government incentives for renewable energy.

Environmental Impact: While hydro energy is generally considered environmentally friendly, it can have some negative impacts, such as altering waterways and affecting local ecosystems. It is important to conduct an environmental impact assessment before installing a hydro system.

The benefits of hydro energy for building decarbonization are clear: it provides a renewable, reliable source of power, reduces a building's carbon footprint, and can lead to significant savings on energy bills. Moreover, by generating their own power, buildings can become more self-sufficient and resilient, which is particularly valuable in the face of increasing energy prices and grid instability.

Evaluating Renewable Energy Impacts

While integral to a low-carbon future, renewable energy systems still incur some environmental and economic costs counterbalancing their sustainability benefits. Responsible scaling requires holistic life cycle analysis and context-specific implementation minimizing negative externalities.

Solar photovoltaics provide clean electricity but require metal mining and intensive manufacturing inputs. Panel disposal and recycling protocols must improve to prevent waste stream impacts. Wind turbines encounter public opposition due to wildlife collisions and aesthetic concerns. Hydropower alters local hydrology through dam construction causing habitat loss. Geothermal projects risk releases of harmful gasses including sulfur compounds.

These challenges seem modest next to fossil fuels' exhaustive toll, but demand scrutiny and innovation. Developers continue advancing technologies and best practices to boost efficiency, durability and closed-loop material flows. Design evolutions shrink components' physical and ecological footprints. Environmental impact monitoring and adaptive management practices also help minimize disturbances.

Cost effectiveness and policy support play pivotal roles for adoption. High renewable energy capital costs deter investment despite long-term savings from avoided fuel expenses and carbon pricing. Multiple financing mechanisms help bridge this gap, including bonds, aggregates and third-party models to distribute burden. Most clean energy infrastructure lasts decades, providing hedge value against fuel volatility.

Project siting strongly determines outcomes, and individual building integration requires rigorous feasibility analysis weighing all tradeoffs. Blanket technology preferences ignore context variability. What sustainable energy portfolio optimally aligns with regional renewable resources, existing infrastructure constraints, climate factors and community priorities provides the key question.

With conscientious implementation tuned to local conditions, renewable energy systems still constitute society's best bet for averting climate catastrophe. But realizing their promise relies on responsible development consciously balancing environmental impacts and community needs against technical and economic potential. Ongoing transparency, accountability and equitable decision-making remain imperative.

The Future of Renewable Energy and Decarbonization

The building sector's renewable energy outlook shines brightly as converging socioeconomic trends propel a clean energy transition. With improved technologies, supportive policies and intensifying climate urgency, structures integrating distributed generation and procuring offsite renewables appear poised for exponential growth.

Ongoing clean energy advances continue driving down financial barriers across solar, wind, geothermal and biogenic solutions. Emerging storage options help overcome intermittent output limitations. Smart meters and load flexibility help align variable generation to dynamic demand profiles. Third-party financing vehicles ease adoption for homeowners and enterprises alike.

Government commitments codifying ambitious carbon reductions by 2030 and 2050 provide tailwinds. Local building codes increasingly mandate or incentivize net zero energy facilities. Clean electricity standards and carbon pricing lift renewable power's competitiveness. Cities lay plans to eliminate fossil fuels from their building stocks within decades.

The transition's speed and equity remain contingent on social and political choices still ahead. While technological and economic trends bode well, deliberate policymaking and incentives must prevent under-resourced communities from being left behind. And the narrow path to 1.5°C warming limitation relies on rapid, across-the-board renewable scaling before the 2030s.

With determined, inclusive action, a renewable-powered building sector future clearly comes into view. The essential technologies already exist to curb emissions through wide scale electrification and biofuel substitution, with costs falling quickly. The greatest variable remains whether society can organize the collective will for broad participation on pace with climate science imperatives. If renewable integration expands exponentially this decade, the building sector may lead the way to global decarbonization.

The Fossil Fuel Transition

The building sector's heavy reliance on fossil fuels for meeting heating, cooling and electrical loads drives considerable greenhouse emissions while incurring substantial health and financial costs. Yet renewable energy technologies promise affordable substitutes offering environmental and social co-benefits. Accelerating this transition critical for climate progress and human wellbeing.

Solar, wind and other renewables increasingly outcompete fossil generation on cost while delivering more resilient and decentralized energy. Electric heat pumps now surpass the efficiency of gas furnaces for space and water heating using a fraction of the energy. Pairing these technologies with improved insulation, passive design and storage unlocks fossil-free operations.

However, upfront costs, inertia and path dependencies inhibit rapid change. Consumer unfamiliarity alongside lack of institutional capacity and financing access deter adoption. But supportive policies and incentives can override these barriers by redirecting investment flows towards sustainability.

The societal impetus for progress quickens as climate consequences intensify. Phasing out building fossil fuel usage provides one of the most substantial and cost-effective means for emissions reductions. The essential technologies and best practices are well understood and improving continually with scale. The growth of renewables remains exponential—what is needed next is mustering the collective action to accelerate this transformation before the narrow window for climate stability shuts.

Sustainable Alternatives to Fossil Fuels

Realizing a fossil-free future for the built environment demands deploying technologies ready to deliver the services buildings require without the emissions baggage. Already mature options for space heating, water heating and appliances carry excellent potential masked behind inertia and path dependency barriers.

Heat pumps employ refrigeration cycles to transfer ambient thermal energy rather than burning fuels. Their unmatched efficiency slashes heating and cooling energy loads by half or more. Mini-split air source systems work in most climates, while ground and water source variants offer additional performance.

Solar thermal installations harness the sun's radiant energy through panels or tubes feeding harvested heat to water tanks or hydronic heating loops. Though regionally variable, solar resources routinely defray 80 percent or more of hotels' and hospitals' water heating demand through these elegantly simple systems.

Electric appliances and HVAC supplant gas-fired counterparts for virtually every building end use. Selecting high efficiency models magnifies emissions reductions from rising renewable penetration of grids. Smart integration of storage and controls further optimizes flexibility.

These technologies shine through compatibility with renewable energy resources and fusion with holistic efficiency strategies. Their modular scalability allows gradual adoption as experience and confidence build. And their lifecycle carbon footprints approach zero as grids green.

Sometimes, the steep upfront costs of purchasing and installing unfamiliar equipment inhibits uptake. Consumer discounts and attractive financing schemes help overcome this adoption barrier. Streamlining permitting and contracting further accelerates the transition.

Fortunately, these technologies mostly pay for themselves through utility savings before equipment lifespan expiry. Additional health and climate co-benefits then continue accruing for years after the payback period. The path towards rapid building stock renewal relies on communicating compelling value propositions around sustainability and economics.

Deep Dive into Heat Pumps

Heat pumps constitute an oft-overlooked pathway for affordable building decarbonization. By leveraging thermodynamic cycles rather than combustion, they supply low-carbon space heating, cooling and water heating with unparalleled efficiency. Initiatives optimizing adoption to displace gas demonstrate profound emissions reduction potential.

These refrigeration-based systems transfer ambient thermal energy rather than burning fuels. Absorbing freely available heat from outdoor air, ground loops or bodies of water allows concentrating energy to useful temperatures for buildings services. The same components reverse flow directions to provide highly efficient space cooling during warmer months.

For every unit of electricity input, state-of-the-art models yield three to five units of building heating. This coefficient of performance (COP) triples or quadruples that of conventional fossil heating. And with building codes now requiring integration of renewable energy, operational emissions keep falling. However, steep upfront investment costs deter wider implementation. Unfamiliarity around reimbursable drilling, piping and installation leaves consumers daunted despite appreciable long term savings. Soft costs like permitting and planning also restrain uptake.

Targeted policy intervention helps address hurdles impeding substitution of gas-fired systems. Utility rebates, tax breaks and preferential financing terms ease adoption. Streamlining and standardization paperwork reduces soft costs. And public education campaigns promote awareness around costs recouped through energy savings. Jurisdictions worldwide now set heat pump deployment targets to meet decarbonization goals. For example the U.S. state of Maine expects to install 100,000 units over this decade. The UK recently announced a plan to roll out 600,000 per year, and China already sees over 30 million annual installations. Ambitious policies and incentives clearly drive aggressive industry growth.

With their distributed flexibility, high efficiency and falling costs, heat pumps constitute an essential and scalable technology for rapid building stock decarbonization. Renewable-powered systems yield little to no carbon emissions while delivering healthier and affordable thermal comfort. The vital next phase involves optimizing supporting policies and financial innovation to unlock wide scale adoption.

Exploring Solar Hot Water Systems

Solar thermal installations for water heating help buildings tap into the emissions-free energy bounty of sunlight. Absorber panels and tubes feed gathered heat to tanks and pipes, displacing conventional water heating fuel combustion. Though deceptively simple, these elegantly direct technologies yield profound carbon and cost savings through prudent implementation.

Two main architectural approaches dominate the growing solar thermal market. Active systems feature sensors, controls and pumps optimizing heat transfer efficiency to storage and distribution loops. Passive designs instead rely on natural convection, trading some performance for simpler installation. Site-specific climate and hot water demand profiles determine ideal configurations.

Uptake worldwide surges as sustainability consciousness swells among consumers and policymakers alike. Solar thermal retrofits now receive robust public rebates from Germany to South Africa to China. Mandates for renewable integration in new construction take effect from Europe to North America and beyond. Some obstacles to universal deployment remain. Solar thermal capacity factors vary widely across seasons and geographies. Upfront hardware and installation costs deter less affluent adopters. While the availability of south-facing exposure limits uptake in dense urban districts. Innovations across manufacturing, materials and information technology aim to tackle these adoption barriers. Emerging polymer absorbers and selective coatings boost efficiency. Creative configurations expand siting flexibility. And smart controls smooth production intermittency through predictive modeling and storage integration.

With conscientious implementation suited to local conditions, solar thermal technology promises to supply a major share of buildings' thermal energy needs. The forthcoming phase of optimization and customization relies on cross-disciplinary creativity and purposeful policy support. But the prospects for emissions reductions remain bright as solar access far outpaces our capacity to harness its potential.

Alternative Refrigerants for Reducing Emissions

Most refrigeration and cooling technologies rely on potent synthetic greenhouse gasses for heat transfer capabilities. However, sustainable alternatives free of global warming impacts now offer direct substitutes. Migrating HVAC&R equipment to these climate-safe coolants remains imperative for mitigating emissions.

Dominant refrigerants like hydrochlorofluorocarbons (HCFCs) were phased down under the Montreal Protocol for depleting stratospheric ozone. Yet most replacements such as hydrofluorocarbons (HFCs) still intensify atmospheric heat capture up to thousands of times more powerfully than carbon dioxide. Even small leaks from aging equipment and piping systems contribute significantly to climate change when aggregated globally. Thankfully, today we have several refrigerant options eliminate direct global warming potential through lower atmospheric impacts:

Hydrofluoroolefins (HFOs): synthesize chemical stability resembling popular HFCs but with radically lowered capacity to absorb infrared radiation as greenhouse gasses. Early HFO variants still carry high upfront costs and marginal efficiency losses.

Hydrocarbons: like propane and isobutane leverage natural refrigeration performance and low l leakage impacts due to their short atmospheric lifespan. Their adoption requires strict safety precautions to contain flammability risks.

Ammonia: an ultra-efficient refrigerant but also introduces toxicity risks requiring careful equipment handling. With prudent precautions and maintenance regimens, it provides unmatched performance economically.

Carbon dioxide: appeals through its ubiquity, lack of direct global warming potential and low cost. The compound requires operating under high pressure with attentive system design to achieve cooling adequacy.

These alternatives demand targeted standards, incentives, training and financing to ease market adoption. But their potential to deliver cooling services without exacerbating atmospheric warming earns them prime status as transitional refrigerants while longer-term technologies progress. Phasing down HFCs this decade remains imperative to bend the emissions curve and propel climate progress.

Transitioning cooling equipment from climate-damaging to climate-safe refrigerants promises enormous collective benefit but faces obstacles around economics and inertia. Technical guidance, financial nudges and regulation can accelerate market transformation to replace lingering hydrofluorocarbons (HFCs). However, some barriers to adoption include:

Cost: Upfront component and retrofit expenses exceed familiar yet inferior HFC systems. Life cycle savings from efficiency gains only accrue over long operating timeframes.

Performance: Capacity, coefficient of performance ratings and operating conditions vary across replacement refrigerants for a given application. Suboptimal selections undermine potential.

Safety: Flammable and toxic alternatives demand vigilant handling, storage and disposal to prevent leaks, explosions and exposure. Fear of liability slows change.

Compliance: Fragmented regulation and weak ambition levels allow continued HFC intimacy. Enforcement resources concentrate elsewhere while environmental harm continues.

Incentives: Rebates, tax breaks and preferential utility rates defray higher incremental first costs to ease purchases. Bulk procurement and contracting smooth adoption at scale.

Education: Extensive training and certification around safety procedures, performance nuances and optimization best practices accelerate competency.

Policy: Functional building codes, refrigerant transition mandates, and manufacturing restrictions pave a path away from relic HFCs towards climate-aligned coolants.

With coordinated action across stakeholders, the cooling sector can phase down antiquated greenhouse gasses in favor of accessible low-global warming potential refrigerants within this pivotal decade. The environmental and economic win-win scenario awaits activation through visionary leadership.

Insulation for Energy Efficiency and Emissions Reduction

Insulation plays a crucial role in enhancing the energy efficiency of buildings and consequently reducing greenhouse gas emissions. This module will delve into the significance of insulation, the types of environmentally friendly insulation materials, and case studies of buildings that have successfully utilized insulation to improve energy efficiency and reduce emissions.

Insulation is a key factor in controlling the flow of heat in and out of a building. It helps maintain a comfortable temperature inside the building, reducing the need for heating or cooling. This reduction in energy use directly translates to lower greenhouse gas emissions, contributing to the decarbonization of buildings.

Insulation works by creating a barrier between different areas of different temperatures. In colder months, insulation prevents the warm air inside the building from escaping to the colder outside environment. Conversely, in warmer months, it stops the hot air outside from entering the cooler interior of the building.

By reducing the demand for heating and cooling systems, insulation significantly lowers energy consumption, leading to a decrease in carbon dioxide and other greenhouse gas emissions associated with energy production. There are several types of insulation materials that are both effective and environmentally friendly. Some of the most common ones include:

Cellulose Insulation: Made from recycled paper products, cellulose insulation is a sustainable and effective option. It has a high R-value (a measure of thermal resistance), making it a good choice for thermal insulation.

Sheep's Wool Insulation: An excellent natural insulator, sheep's wool can absorb and release moisture without compromising its insulating properties. It is renewable, biodegradable, and requires less energy to manufacture than most synthetic insulations.

Cork Insulation: Made from the bark of cork oak trees, cork insulation is another sustainable choice. It is renewable, recyclable, and has good insulating properties.

The Role of Policy and Regulation

The use of fossil fuels in buildings is currently regulated by a variety of policies at the international, national, and local levels. Policy frameworks steering the building sector's sustainability transformation encompass a multifaceted web of mandates, incentives and disclosure mechanisms across governance levels. While codes and standards dominate for new constructions, addressing the far larger existing stock requires nuanced regulation easing transitions without overburdening impacted groups.

Prescriptive building codes now commonly require minimum energy efficiency, electrification readiness and renewable energy integration in major renovation and new build projects. Performance-based alternatives allow wider design flexibility to achieve footprint reductions. However, codes primarily target the narrow subset of new annual additions, leaving the vast majority of structures unmodified.

Overcoming status quo bias relies on carrots more than sticks, at least initially. Financial incentives like tax breaks, preferential utility rates and rebates help owners internalize societal decarbonization benefits. Municipal ordinances simplifying permitting and contracting lower soft costs. And voluntary standards recognizing low-carbon leadership beyond legal minima build market experience.

Yet tailored policy best serves equitable outcomes. Exemptions and graduated phase-ins ease burdens for lower-resourced cohorts like affordable housing providers. Disclosure mandates tracking transparency, rather than outright bans, allow change without immediately jeopardizing operations. And support for workforce training, public housing upgrades and clean energy access accompanies regulation.

Ultimately, coherent policy signaling unifies around consistent messaging and feasible timelines recognizable across sub-jurisdictions. With social outcomes directing financial flows, regulated and voluntary action can symbiotically reinvent buildings for a just, resilient future.

Policies and Regulations Related to Building Decarbonization

Building decarbonization is not just a technical challenge but also a policy issue. Policies and regulations play a crucial role in driving the decarbonization of buildings. Policies set the rules of the game by providing incentives for good and sustainable practices and penalties for unsustainable practices. Governments around the world have recognized the need to reduce the carbon footprint of buildings and have implemented a variety of policies and regulations to encourage and enforce decarbonization efforts.

By setting minimum standards for energy efficiency and carbon emissions, building codes and standards ensure that all new buildings and major renovations contribute to decarbonization efforts. They also drive innovation in the building industry, as architects, engineers, and builders seek to design and construct buildings that not only meet but exceed these standards.

Financial incentives make it more attractive for building owners to invest in energy efficiency and renewable energy, reducing the payback period for these investments. Mandatory reporting and disclosure increase transparency and accountability, encouraging building owners to reduce their energy use and carbon emissions. These policies and regulations can take various forms, including:

Building codes and standards: These set minimum requirements for energy efficiency and the use of renewable energy in new buildings and major renovations. They may also include requirements for the use of low-carbon building materials and construction methods.

Energy performance certificates: These are mandatory in many countries for all buildings that are sold or rented, providing information on a building's energy use and carbon emissions.

Financial incentives: These can include grants, tax credits, and low-interest loans for energy-efficient and low-carbon buildings.

Mandatory reporting and disclosure: Some jurisdictions require building owners to regularly report on their building's energy use and carbon emissions, and to disclose this information to potential buyers or tenants.

Financing and Policies for Building Decarbonization

Achieving absolute decarbonization requires radical, rapid reimagination of the built environment through investments in retrofits, electrification and distributed energy resources. This outlay relies on unlocking vast pools of public and private capital steered towards unambiguous emissions cuts, resilience and affordability. Innovations in financial engineering balance risk, return and impact while transparency principles check outcomes.

Public funds like government bonds, multilateral loans and municipal contracts seed initial demonstrable pilots validating technical and economic viability at scale. Resultant cost and performance data informs policy design adjustments and de-risks subsequent private deployment. Philanthropic grants also support critical planning and advocacy functions around equity.

Meanwhile financial sector leaders respond through ethical allocation alignment with climate imperatives and social responsibility. Mainstream green bonds, loans and sustainability-linked instruments funnel institutional investment into vetted low-carbon building projects with lending terms rewarding verified progress. Platforms bundling improvements across property portfolios achieve outsized impact through standardization.

But accountability principles guide the proliferation of instruments – transparency requirements stress comprehensive tracking standards integrated into deals ex-ante while industry coalitions share best practices and guard against greenwashing. Impact metrics assess multidimensional criteria including affordability, resilience and community health alongside avoided emissions and ecological sensitivity. With collaboration and innovation, sufficiently massive capital now mobilizes towards equitable building decarbonization. The essential next phase involves steadfast leadership and cooperation to enable rapid, whole-system transformation benefitting all through regenerative design.

Financing Options for Building Decarbonization

Funding building decarbonization relies on mobilizing vast capital flows from diverse public and private sources towards technologies and designs delivering absolute carbon cuts. These complementary funding pools each contribute unique strengths around risk tolerance, impact focus and site specificity necessary for comprehensive transformation.

Public financing instruments like government bonds, multilateral loans and municipal contracts provide substantial seed capital for demonstration projects validating technical potential and business models. Resultant cost and performance data informs adjusted policy support for subsequent mass deployment. Grants also back critical sustainability planning and advocacy functions.

Meanwhile private finance increasingly views climate-aligned real assets as ethical long-term stores of value with political tailwinds. Mainstream green bonds, loans and sustainability-linked instruments allow institutional investors to fund portfolios of low-carbon buildings and earn income from energy savings. Green banks, Pipe funds and crowdfunding platforms also finance discrete distributed projects.

But principles of transparency and accountability now permeate investment decisions to check outcomes against intentions. Reporting frameworks like the EU sustainable finance taxonomy stress comprehensive impact assessment qualified by science-based targets. This scrutiny drives innovation in what constitutes "green" buildings amidst competing definitions.

Ultimately, the mechanisms of finance must serve equitable decarbonization and resilience above all else. With historically siloed funding flows now intersecting around aligned values and unprecedented collective need, the built environment can lead whole economy regeneration.

Incentives for Building Decarbonization

While building decarbonization demands substantial capital investment, incentives help ease transitional barriers around unfamiliarity and constrained resources. Financial nudges like tax relief, rebates and preferential utility rates reward owners internalizing social emissions reduction benefits through property upgrades. Code allowances also lower compliance costs for voluntary leadership.

Tax credits reduce tax burdens for owners installing efficiency measures, electrification systems and distributed renewables. These help address upfront capital barriers beyond value recouped solely from utility savings over time.

Enhanced federal credits targeting commercial properties would impact carbon-intensive existing buildings.

Rebates directly subsidize incremental material and installation costs to enhance appeal. Programs reimbursing incremental heat pump water heater price premiums over conventional resistance models accelerate retirement of legacy equipment. Conditional discounts for permitted deep efficiency retrofits drive net zero outcomes.

Expedited approvals, graduated fee structures and design flexibility build on intrinsic motivations for beyond-code design. Allowing renewable appendages qualifying as standard repairs speeds adoption. Partnerships with utilities to smooth new load interconnection reduce transaction costs.

Yet incentives should balance prescriptiveness with flexibility for holistic building-specific pathways. They must also stress equity for under-resourced groups through tiered access and workforce development. With social priorities guiding investment, financial support and allowances can profoundly impact decarbonization.

Beyond tax relief, direct public co-funding through rebates and grants drastically reduces owners' cash outlays for sustainability upgrades. These further tip the scales for carbon-conscious retrofit decisions yielding public health and emissions savings over time. Customized support across technology categories factors into a holistic context.

Rebates directly subsidize the incremental price premiums of energy efficient equipment compared to conventional alternatives. Discounts rewarding heat pump adoption and strategic electrification acknowledge marketplace obstacles for nascent solutions. Compensating a share of weatherization costs addresses split owner-tenant motivations.

Grants allocated through programs like the Department of Energy's State Energy Program fund regional demonstrations establishing best practices for technology deployment and project management. Competitive solicitations target under-addressed market segments facing capital access barriers. Workforce training grants also equip displaced fossil fuel laborers with retooled skill sets.

Means-tested grant allocations through agencies like NYSERDA reserve the highest co-funding rates for lower-income households via partnerships with local contractors. Funds administered by utilities for income-qualified community solar subscriptions unlock clean, resilient power for renters.

With social outcomes prioritized in funding decisions, customized public co-capital enables equitable building decarbonization progress at decisively early and visible scales. Demonstrating potential builds momentum for broader action.

Policies Promoting Building Decarbonization

Policy frameworks steering building decarbonization integrate mandatory codes, voluntary standards and public education efforts towards shared emissions reduction targets. Regulation establishes minimum compliance floors while incentives encourage beyond-code leadership. Holistic transformation relies on cooperation among government, industry and communities.

Building energy codes now commonly mandate minimum renewable integration, electrification readiness and performance efficiency - new California standards require rooftop solar on most new homes. Requirements grow more stringent over 3-6 year adjustment cycles informed by feasibility analyses. Cities like New York and Toronto set their own codes exceeding state or provincial baselines.

Beyond compliance carrots, demonstration rewards through expedited permitting, design flexibility and graduated fee structures validate and scale replicable models. Dynamic codes and outcomes-based alternatives also promote tailored, net-zero pathways. Voluntary rating schemes like LEED, Living Building Challenge, and the International WELL Building Institute accelerate market education.

But equitable outcomes direct policy conversations. Exemptions protect lower-resourced communities during transitional phases. Scaling training, funding efficiency upgrades for public housing and expanding community solar subscriptions prevent underserved groups from being left behind in the clean energy economy.

With social priorities guiding requirements and incentives, regulation can reposition buildings as sites of health, resilience and sustainability abundantly accessible to all. Ongoing community input and cooperation focus policy implementation on justice.

Navigating the Financial and Regulatory Landscape

Unlocking low-carbon buildings at scale relies on optimizing investments within intricate policy environments prime for disruption. Strategic navigation serves social and climate priorities when grounded in cooperation, demonstration and social outcomes.

A myriad of financing mechanisms now advance climate-aligned real assets, but project teams must identify options suiting needs. Governments offer bonds, loans and contracts seeding demonstrable pilots to de-risk innovation. Mainstream green financial instruments allow bundling improvements across portfolios to access institutional investment. Distributed networks like Pipe funds and crowdfunding platforms finance community-rooted initiatives.

Success follows comprehensive life cycle cost-benefit analysis and sound business cases conveying sustainability benefits - from emissions and ecological sensitivity to wellbeing and resilience. Partnerships with councils, syndicates and ESCOs build capacity while ensuring local participation.

Expanding codes mandate minimum on-site renewable shares, electric readiness and efficiency upgrades for most new buildings and major renovations - with cities exceeding national or state baselines. Dynamic outcomes-based compliance provides tailored net zero pathways. Voluntary programs like LEED educate markets on best practices.

Effective teams integrate code compliance from project inception through post-occupancy while monitoring policy developments. Early authority engagement clarifies interpretations and codes-in-development to prevent misalignment. Multi-stakeholder collaboration addresses complex challenges by balancing priorities.

Incentives ease transitional obstacles for lower-resourced communities until governments implement economy-wide measures. Job training programs and public housing retrofits ensure decarbonization affords economic mobility. Community benefit agreements guarantee affordable access and participation roles to prevent exclusion.

By centering social priorities, financing and regulation can transform buildings equitably. Innovation guided by justice future proofs communities through ethical regeneration. Ongoing cooperation and demonstrations builds this future.

The Future of Financing and Policies for Building Decarbonization

With buildings responsible for over one-quarter of greenhouse emissions, the sector's swift decarbonization grows more pressing yet viable through emerging policy and finance innovations. As governments mandate absolute carbon cuts, ethical investment movements channel unprecedented capital flows towards equitable climate solutions centered on justice.

Building codes now commonly require minimum renewable shares, electric readiness and efficiency upgrades for most new constructions and major renovations. Codes tighten on regular adjustment cycles as local ordinances push ambitious timelines. Outcomes-based compliance alternatives also promote customized net zero pathways.

Meanwhile loan underwriting processes increasingly quantify sustainability risks, directing investment towards resilient properties. Mainstream green bonds and loans finance portfolios of upgrades bundled for scale using distributed ledgers. Municipal green banks, crowdfunding platforms and new securitization vehicles draw diverse funding sources.

But balanced policies centering social outcomes guide progress. Carbon pricing and benchmarking schemes initially exempt vulnerable communities before later emulating explicit price signals. Taskforces shape community benefit agreements ensuring affordable housing access and clean energy jobs accompany development.

Ongoing community participation focuses policy implementation and stewardship roles for impacted groups. Demonstrations validating co-benefits counter misconceptions through experiential education. And transparent multidimensional impact data informs regular, localized adjustments.

The resulting integration of investment, regulation and social priorities builds towards just outcomes benefitting all. With innovation directed towards shared values, the unprecedented systemic disruption underway can equitably reinvent buildings for a collective future.

    1.

As this analysis has explored, buildings offer immense potential for driving rapid decarbonization through integrated efficiency, electrification, and renewable energy adoption. Tackling their vast direct emissions and those enabled across supply chains constitutes a climate imperative and economic opportunity. Viable solutions already exist - from heat pumps to solar modules to alternative refrigerants and beyond. While upfront costs pose barriers, financial innovations like preferential loans, incentives and ethical investment vehicles help ease the transition. Codes and standards will continue ratcheting ambitions upwards through transparent planning processes centering equity.Ultimately, securing a stable climate future requires activating known technologies and policies towards exponential scale guided by justice. Buildings sit at the nexus of climate action and human service provision, embedded in communities and daily life. Transforming their systems and spaces promises cascading ripples advancing both sustainability and wellbeing. Through purposeful leadership and unprecedented cooperation around our shared zero-carbon objectives, the built environment can help lead whole economy regeneration to benefit all.

From Thought to Action: Real World Examples

The Bullitt Center

Seattle, Washington
The Bullitt Center sets high standards for energy-efficient commercial architecture to maximize sustainability. Major features include:

- Thick highly-insulated walls and triple-glazed windows that greatly reduce heat loss in winter and heat gain in summer, improving efficiency.
- Use of cutting-edge energy-efficient appliances and LED lighting throughout to substantially lower electricity consumption.
- A state-of-the-art HVAC system and solar water heating to minimize energy required for climate control and hot water needs.
- An extensive rooftop solar panel array that enables the building to operate as a net-zero energy facility by generating as much renewable power as it consumes.

With its heavy investments in envelope efficiency, high-performance equipment, and on-site solar energy production, The Bullitt Center stands out for self-sufficient, sustainable design. More information can be found at
https://bullittcenter.org

The Edge

The Edge office building in Amsterdam sets high standards for smart, efficient commercial design through features like:

- An intelligent lighting system with sensors that adapts illumination in real-time based on occupancy and natural light to minimize electricity use.
- An architectural design that maximizes natural light and air circulation for efficiency. It uses a variable refrigerant flow (VRF) HVAC system to dynamically adjust cooling/heating needs.
- Advanced energy management via a sophisticated Building Management System (BMS) that optimizes energy use across functions in real-time.

With its cutting-edge lighting, HVAC, and energy management systems tailored for efficiency, The Edge exemplifies intelligent, optimized sustainable architecture. More can be found at:https://edge.tech/developments/the-edge

The Crystal

The Crystal building sets high standards for sustainable commercial architecture in London, Ontario through features such as:

- Integration of photovoltaic solar panels and high-performance insulation to minimize energy loss.
- Solar thermal panels to provide the building's hot water needs with renewable energy.
- Efficient ground source heat pumps for heating/cooling using stable underground temperatures.
- An emphasis on indoor environmental quality and abundant natural daylight to reduce artificial lighting needs.

With its renewable energy production, envelope efficiency improvements, passive design, and efficient HVAC system, The Crystal excels in high-performance sustainable architecture. More details can be found at: https://www.inawe.in/wp-content/uploads/2015/12/The-Crystal-Sustainability-Features.pdf

TELUS Garden

Vancouver, Canada:

The TELUS Garden development in Vancouver demonstrates innovative sustainability through features such as:

- A district energy system that recovers waste energy from cooling processes to heat the building.
- A rainwater harvesting system and rooftop solar array contributing to water and energy efficiency.

By capturing energy and water streams that typically would be lost, TELUS Garden exemplifies cutting-edge systems thinking applied to sustainable commercial architecture. More details available at: https://telusgarden.com/sustainability/

Sparta Clinic Medical Facility

Sparta, Wi

Gundersen Health System's Envision Sustainability Program aims to pursue energy efficiency through low-energy building design coupled with renewable energy. A prime example is their 35,000-square-foot Sparta Medical Clinic in Sparta, Wisconsin which opened in January 2017.

The clinic was built with a variety of sustainability features to minimize energy usage, with a goal of only 35 kBtu/sq. ft. This is 50% less than an average clinic's energy consumption. After a year of operation, the clinic has already exceeded expectations, using only 33 kBtu/sq. ft.

The clinic integrates both on-site and off-site solar energy. It has a 100 KW rooftop solar photovoltaic array. Additionally, 220 KW of solar energy is purchased from a nearby community solar project. Together these solar sources generate enough renewable power to make the building energy independent.

Geothermal wells provide heating and cooling. Forty 300 foot deep wells connect to a distributed geothermal heat pump system, lowering energy use by 15 kBtu/sq. ft.

Other efficiency features include:

- A decentralized heat pump system
- A heat recovery ventilation system
- Spray foam insulation
- Thermally broken frames
- Double pane windows
- Interior/exterior LED lighting with controls
- Advanced HVAC control systems
- IT power management

In the first year, these measures have achieved over $68,500 in energy savings for Gundersen. The clinic demonstrates how to combine efficiency with on-site and off-site renewable energy to create a high-performance sustainable healthcare facility. More information is available at:
https://betterbuildingssolutioncenter.energy.gov/node/7906/pdf#:~:text=Sparta%20Clinic%20was%20built%20with,open%20just%20over%20a%20year.

St. Peter's Multi-Family Affordable Housing

New Orleans, LA

Hurricane Katrina inspired the founding of disaster recovery nonprofit SBP (originally St. Bernard Project) in 2005. By 2018, they had rebuilt 1,680 disaster-impacted houses nationally. Recently SBP has focused on affordable, resilient housing, exemplified by their first multifamily project - the St. Peter Apartments in New Orleans.

The 3-story, 50-unit St. Peter opened in 2020 in the gentrifying Mid-City neighborhood. 29 units are affordable at 60% AMI, with subsidies. 21 units are market-rate up to 120% AMI. Half the units prioritize veterans. The project provides much-needed affordable housing to protect lower income residents from potential displacement.

The St. Peter is Louisiana's first net zero multifamily building. 450 rooftop solar panels and on-site battery storage enable net zero performance. The system can island from the grid to supply backup power during disasters. Passive design, energy-efficient technologies, and low-carbon materials also contribute.

SBP tracks energy use through a master meter and 50 individual apartment meters. Their staff reviews consumption data with residents to further reduce demand. Keeping utility bills low and consistent helps residents budget limited incomes.

Financing stacked low-income housing tax credits, loans, utility and nonprofit partnerships. Construction engaged AmeriCorps volunteers to reduce costs 10-40% versus contractors alone.

The St. Peter exemplifies SBP's strategy to provide affordable, resilient housing that allows vulnerable populations to remain in place through neighborhood transitions. Additional projects are now being pursued locally and nationally through their Opportunity Housing Program. Additional information is available at: https://www.huduser.gov/portal/casestudies/study-040121.html#:~:text=Reduces%20Energy%20Costs-,The%20St.,attention%20to%20residents%27%20energy%20use.

Heat Pump Systems

Toronto, Canada:

The City of Toronto has implemented various innovative sustainable retrofits including:

- Municipal Buildings retrofitted a Ground Source Heat Pump (GSHP) system to significantly improve energy efficiency.
- A condo retrofit replaced an old chiller with air-to-water heat pumps and reduced gas boiler heating to realize major energy savings.
- A 1987 townhouse complex retrofit installed cold climate air source heat pumps (CC-ASHP) targeting a substantial drop in overall energy use.

These examples demonstrate Toronto's commitment to improving the efficiency of existing buildings through cutting-edge heat pump and renewable energy systems. More available at:

https://www.toronto.ca/wp-content/uploads/2017/11/915c-City-of-Toronto-Energy-Conservation-Demand-Management-Plan-2014-2019-Section-1.pdf

Solar Thermal Installations:

In 2018, the Debakey VA Hospital in Houston partnered with Sunshine Plus Solar to install a large solar thermal system to provide sustainable hot water for veterans' rehab pools.

- 40 SunEarth Empire Chrome flat plate collectors were mounted, sized to meet 80% of the medical center's annual hot water demand. This allows the hospital to leverage renewable energy for the majority of its veterans' therapy pool heating needs year-round.
- It marked a major milestone for the Debakey VA, significantly advancing their sustainability commitments while supporting essential veteran healthcare services. The project demonstrates the efficacy and value of large-scale solar thermal technology implementations at medical facilities.
- SunEarth solar collectors and Sunshine Plus Solar installation were key.

Further project details and contacts:

SunEarth: www.sunearthinc.com

Sunshine Plus Solar: https://sunshineplussolar.com/

Incentives

Better Buildings Challenge: This initiative aims to improve the efficiency of American buildings by at least 20% over 10 years, with achievements including significant energy and water savings, financial benefits, and widespread partner participation.

Clean Energy for Low Income Communities Accelerator (CELICA): Focused on reducing energy bills for low-income communities, CELICA has committed substantial resources to enhancing energy efficiency and renewable energy benefits in these areas.

About the Author

Ryan Kmetz is an experienced Climate Change, Sustainability, and Resilience professional dedicated to investigating the pressing environmental issues of today and identifying sustainable solutions for tomorrow. Ryan's passion for the natural world and fascination with infrastructure has driven his multifaceted career advancing climate resilience across diverse regions and organizations.

Leveraging expertise in geographic information systems (GIS) and data science, Ryan works cross-collaboratively to engage stakeholders and drive impact. His analytical approach combines data synthesis, technology integration, and geospatial mapping to provide actionable insights tailored to local and regional climate challenges. Over the past decade, Ryan has worked extensively with state municipalities, city governments, college campuses, and nonprofits helping develop comprehensive sustainability plans and climate action strategies.

Ryan draws on his diverse educational background in pursuing his passion for sustainability. He holds a Master of Science in Environmental Studies from Antioch University New England, where he researched best practices using ecological footprint analysis for industry. Additionally, Ryan earned a Bachelor of Science in Biology from Le Moyne College. He is a certified APPA Educational Facilities Professional (CEFP) and an Institute for Sustainable Infrastructure Envision Sustainability Professional (ENV SP).

Ryan builds stakeholder buy-in through data-driven insights combined with compelling narratives around climate justice and sustainable community development. He continues exploring new technologies and collaborative opportunities to drive meaningful climate resilience outcomes across society. In his spare time, you can find Ryan enjoying time with his family and friends. He is an avid baseball fan and decent woodworker.